Math Made Simple:
First Grade Unit 1
Dr. Jannell Pearson-Campbell

Copyright @ 2024. Dr. Jannell D. Pearson-Campbell

Imprint: Independently published
lAll Rights Reserved

✨ **Why Math Made Simple?**
- Easy-to-follow worksheets designed for First Grade students.
- Fun and colorful exercises to keep your child engaged and motivated.
- Reinforces important math skills like number recognition, counting, addition, and more.
- Perfect for at-home learning, homeschooling, or extra practice alongside schoolwork.

📚 Print, Learn, and Succeed – it's that simple

Visit

drjpcdownloads.etsy.com for printable math worksheets

keep it SIMPLE

Create your own Notebook
Use Markers or Colored Pens

Explain your thoughts using a dry erase markers

Use Manipulatives for Addition, subtraction, Multiplication and Division

Math Made Simple :

Use a compatible device, such as a laptop or tablet.

Connect the device to a projector or smartboard.

Utilize screen-sharing options for easy projection of the material.

Create an interative lesson using a dry erase board

Explore more resources at drjpcdownloads.etsy.com and on Amazon by Dr. Jannell Pearson-Campbell!

Unit 1	Page Numbers
Counting from 1-20	7-50
Counting from 1-100	51-86
Addition Problems	87-102
Subtraction Problems	103-114
Place Value Problems	115-124

COUNTING FROM 1-20

Understanding the Number 0

In Words	Symbol	Count
Zero	0	

Now It's Your Turn

In Words	Symbol	Count
Zero	⓪	
Zero	⓪	

Understanding the Number

1

In Words	Symbol	Count
One	1	

Now It's Your Turn

In Words	Symbol	Count
One	1	🐸
One	1 (dotted)	

TRACING NUMBERS

Count the item and trace the number word.

Trace the number.

Understanding the number

2

In Words	Symbol	Count
Two	2	

Not it's *Your* Turn!

In Words	Symbol	Count
Two	2	🐸🐸
Two	2	

TRACING NUMBERS

Count the items and trace the number word.

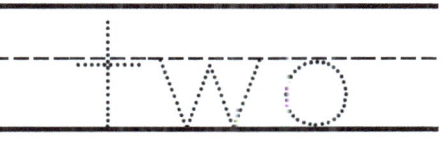

Trace the number.

Understanding the number 3

In Words	Symbol	Count
Three	3	

Not it's *Your* Turn!

In Words	Symbol	Count
Three	3	🐸🐸🐸 (1, 2, 3)
Three	3	

TRACING NUMBERS

Count the items and trace the number word.

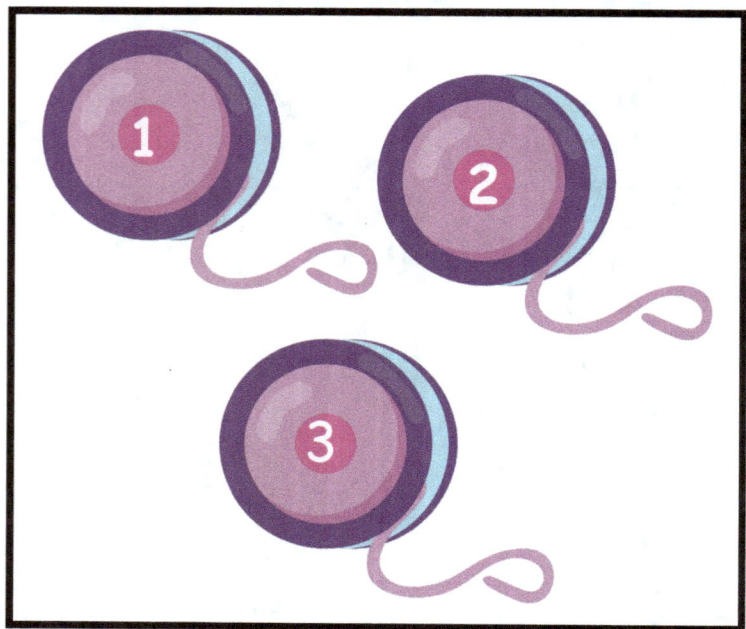

Trace the number.

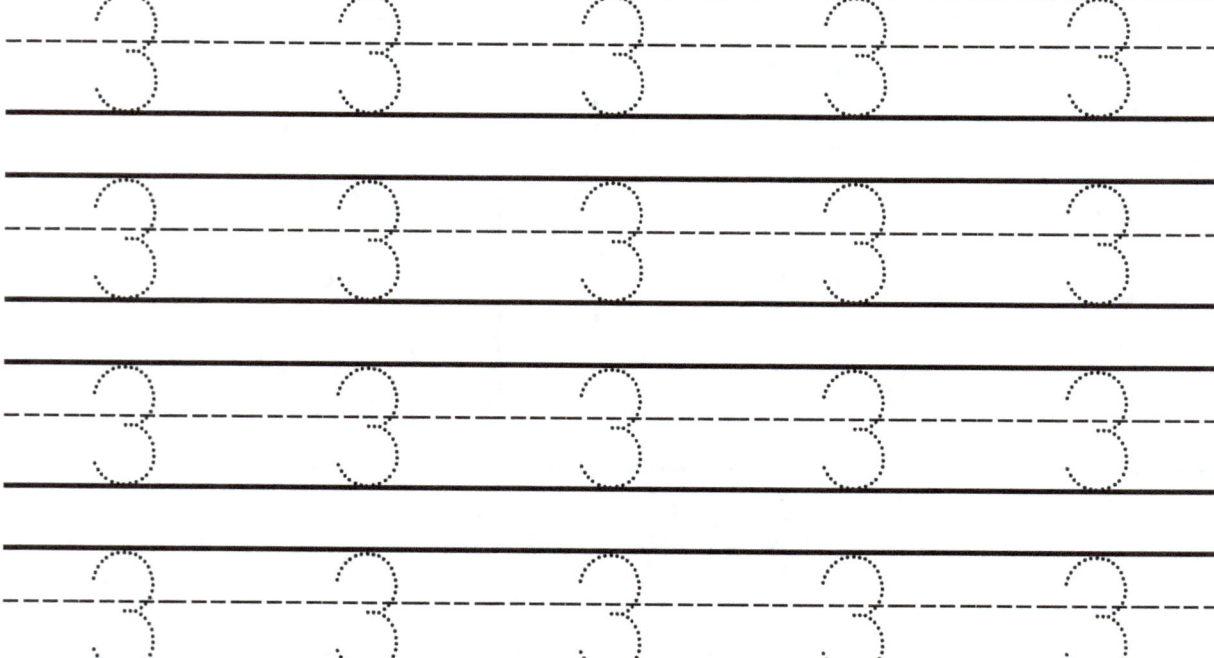

Understanding the number 4

In Words	Symbol	Count
Four	4	

Not it's *Your* Turn!

In Words	Symbol	Count
Four	4 (dotted)	🐸1 🐸2 🐸3 🐸4
Four	4 (dotted)	

TRACING NUMBERS

Count the items and trace the number word.

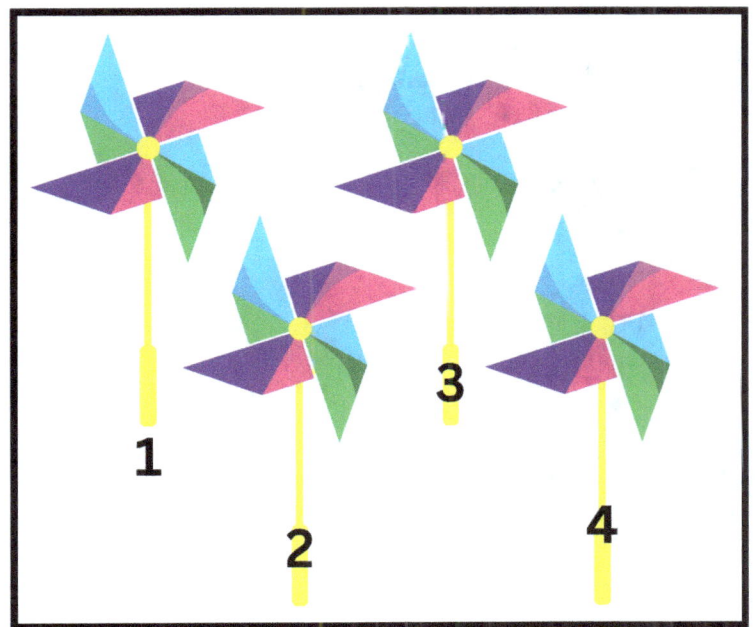

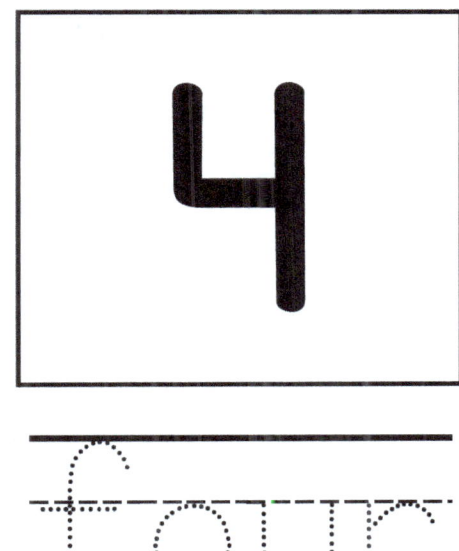

four

Trace the number.

Understanding the number 5

In Words	Symbol	Count
Five	5	

Not it's *Your* Turn!

TRACING NUMBERS

Count the items and trace the number word.

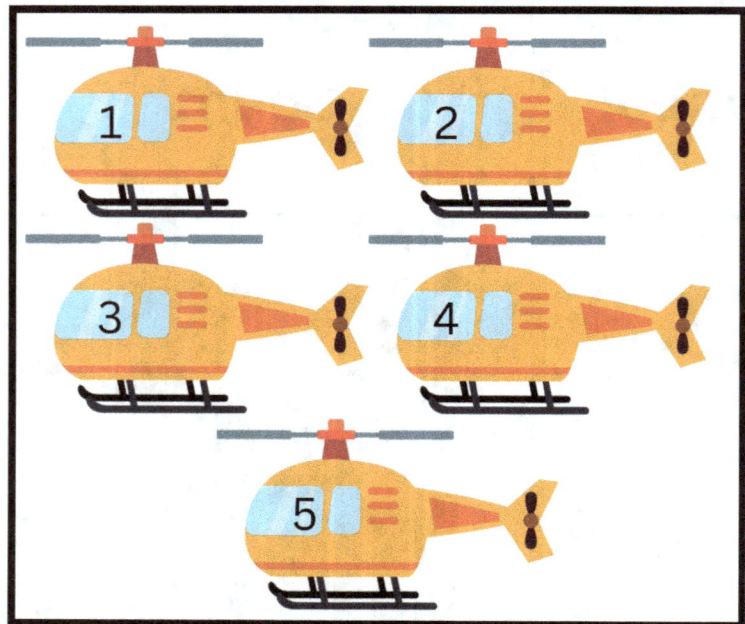

Trace the number.

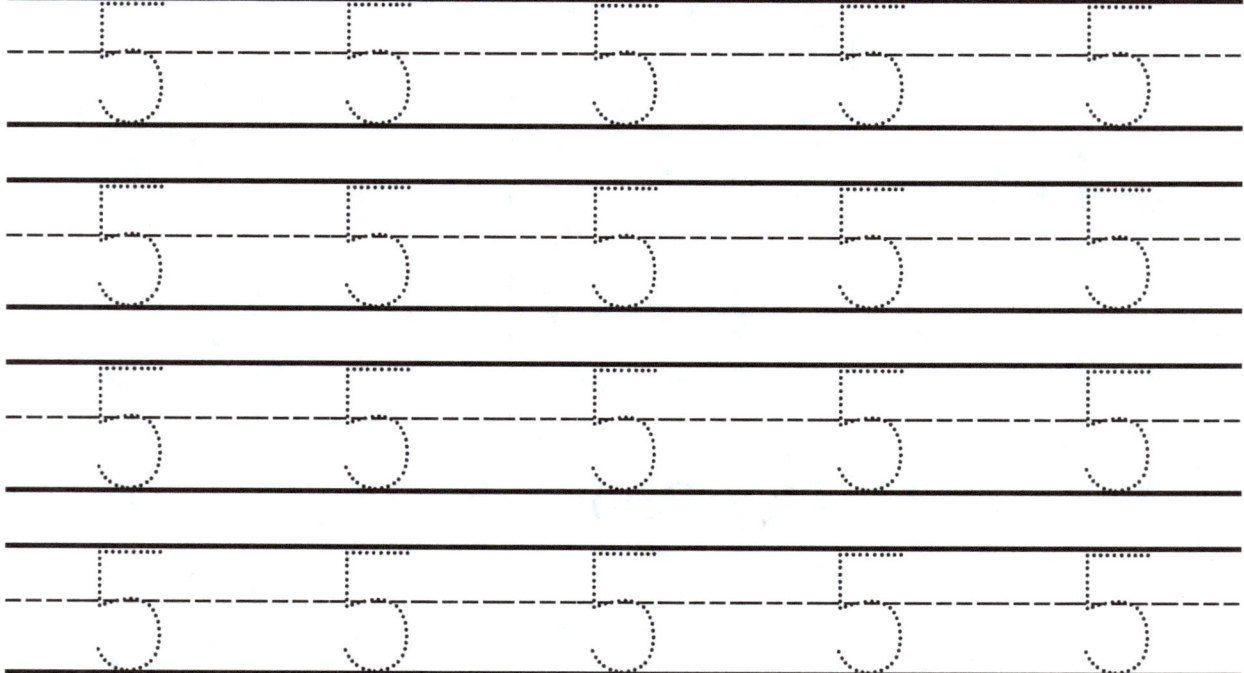

Understanding the number 6

In Words	Symbol	Count
Six	6	1 2 3 4 5 6

Not it's *Your* Turn!

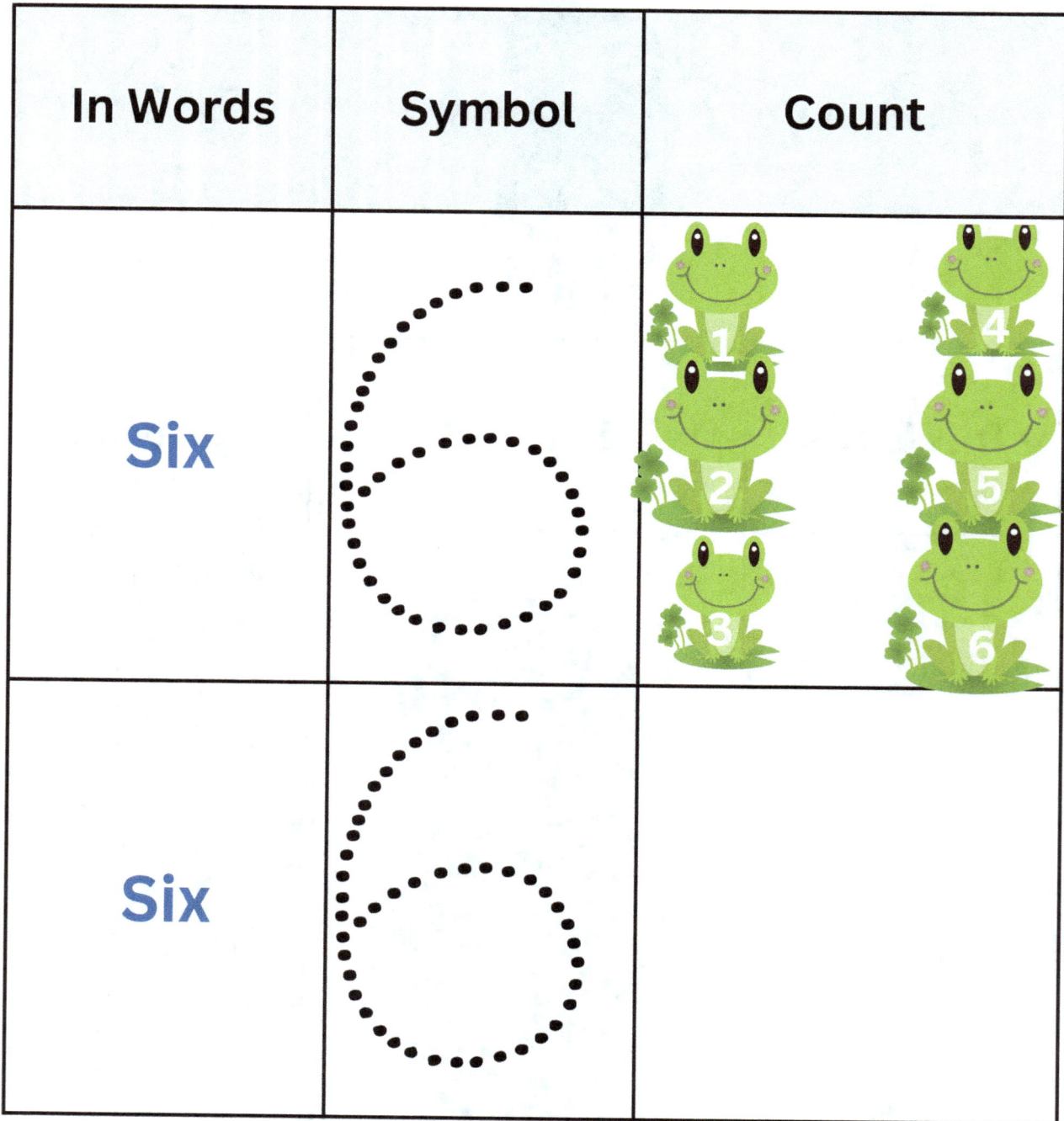

TRACING NUMBERS

Count the items and trace the number word.

Trace the number.

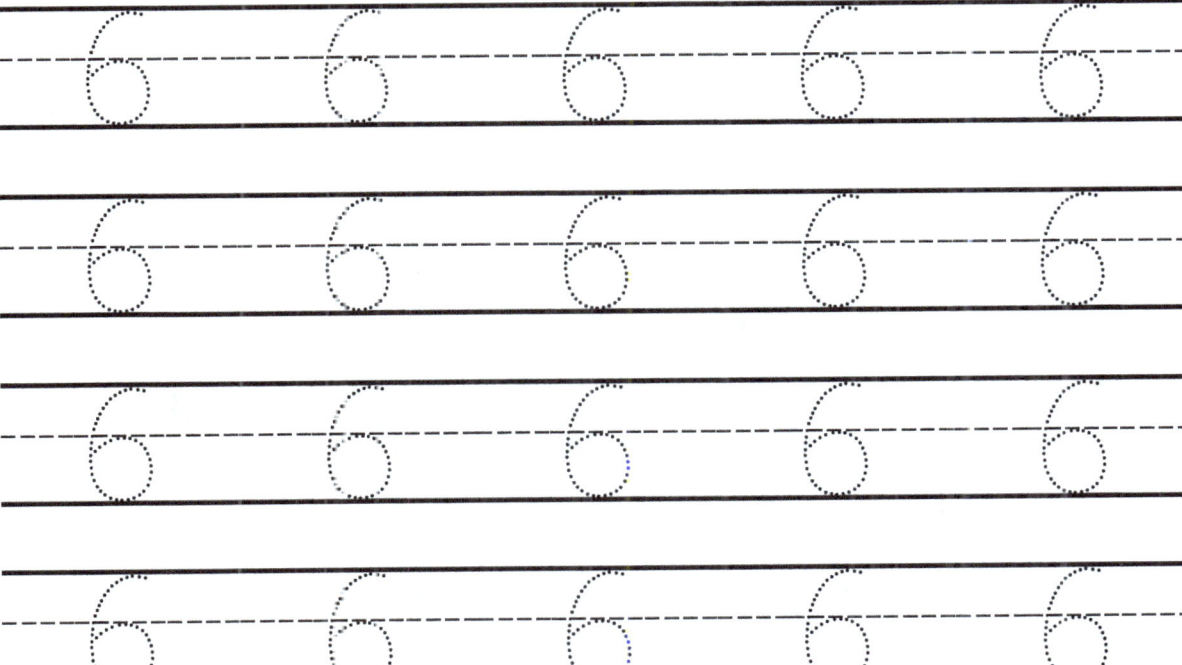

Understanding the number 7

In Words	Symbol	Count
Seven		1 2 3 4 5 6 7 (frogs)

Not it's *Your* Turn!

In Words	Symbol	Count
Seven	7	🐸1 🐸2 🐸3 🐸4 🐸5 🐸6 🐸7
Seven	7	

TRACING NUMBERS

Count the items and trace the number word.

Trace the number.

Understanding the number 8

In Words	Symbol	Count
Eight		1 2 3 4 5 6 7 8 (frogs)

Not it's Your Turn!

In Words	Symbol	Count
Eight	8	1 2 3 4 5 6 7 8
Eight	8	

TRACING NUMBERS

Count the items and trace the number word.

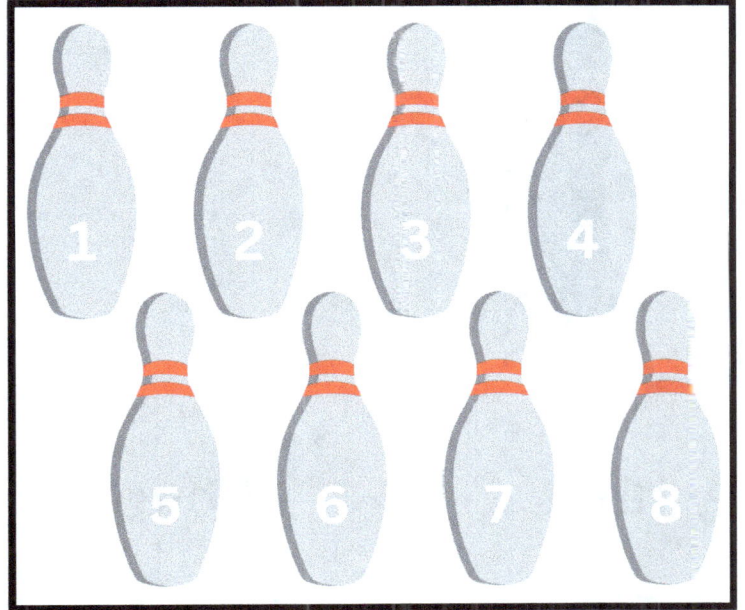

Trace the number.

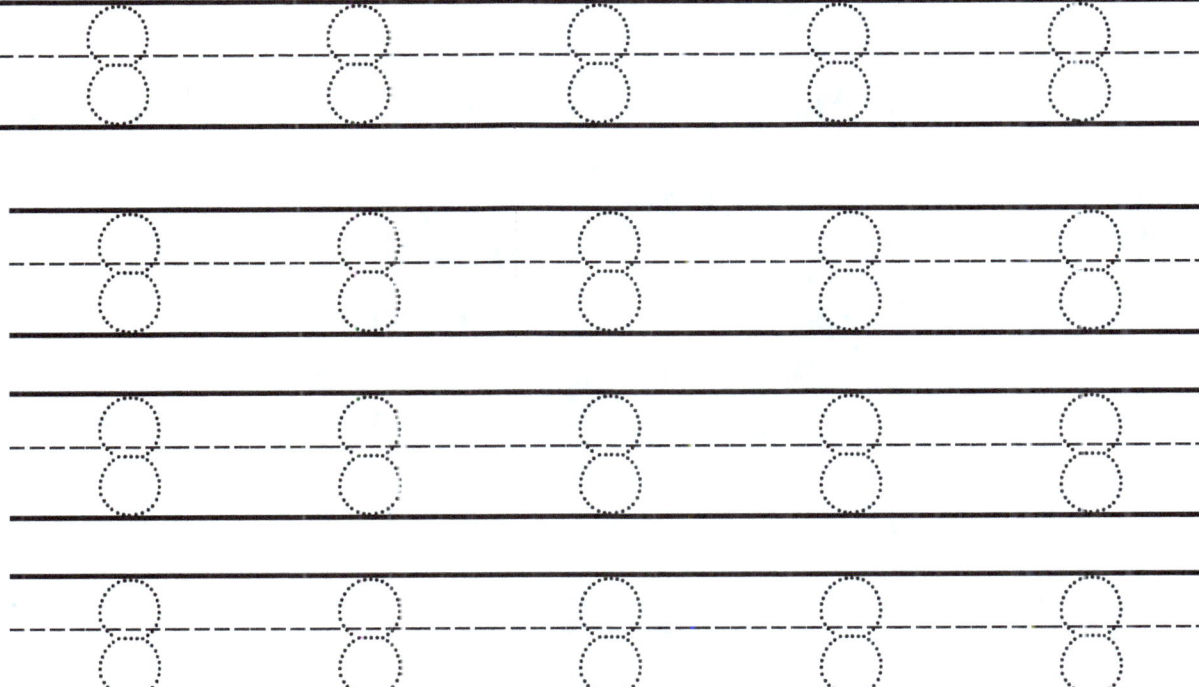

Understanding the number 9

In Words	Symbol	Count
Nine	9	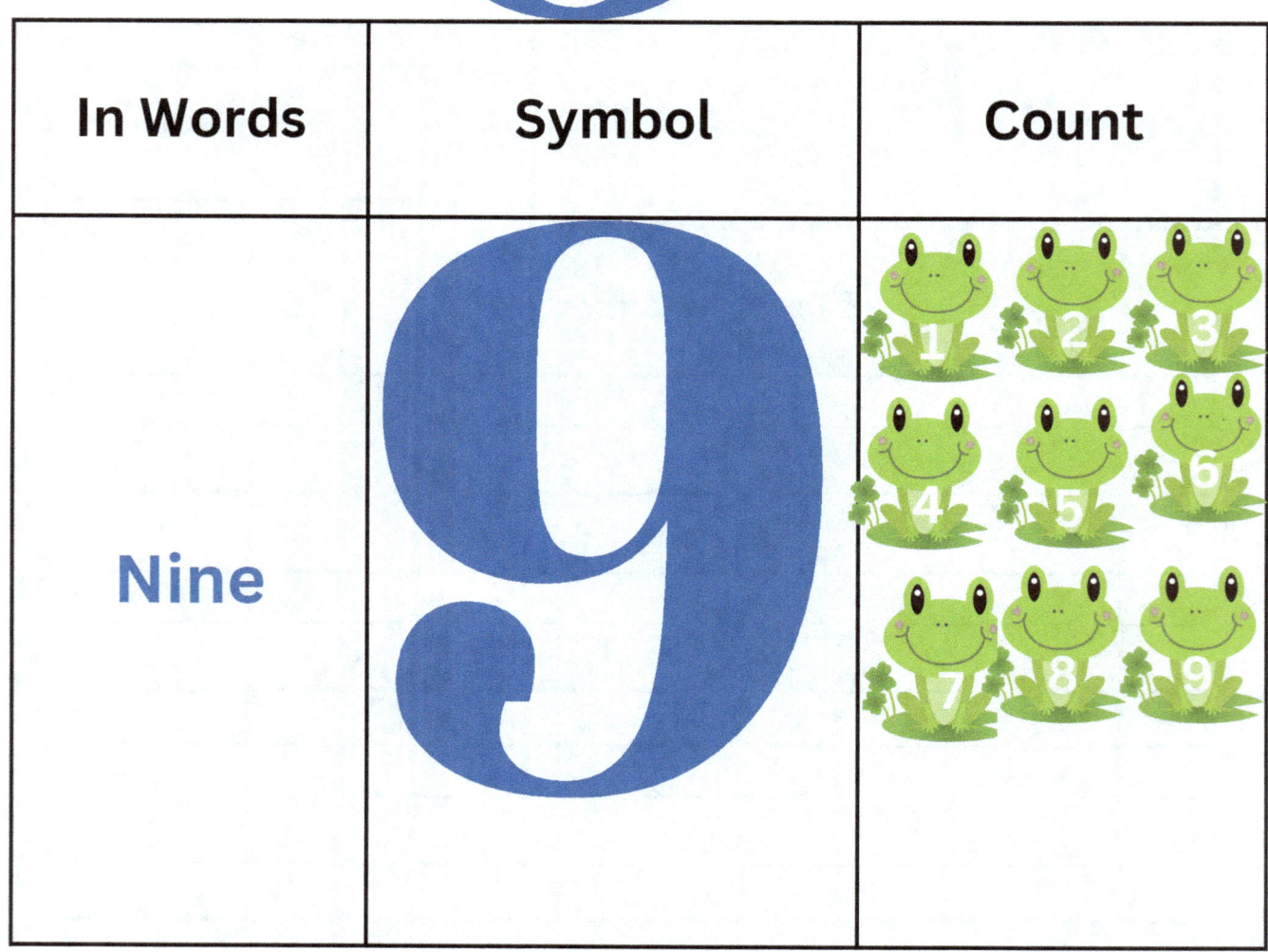

Not it's *Your* Turn!

In Words	Symbol	Count
Nine	9	1 2 3 4 5 6 7 8 9
Nine	9	

TRACING NUMBERS

Count the items and trace the number word.

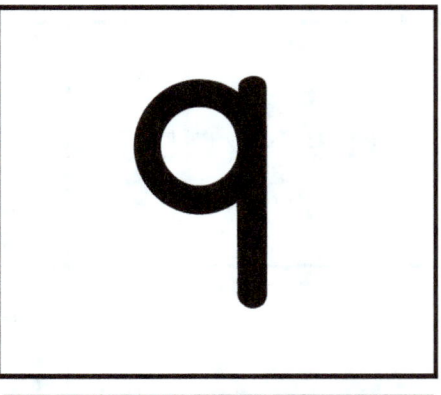

Trace the number.

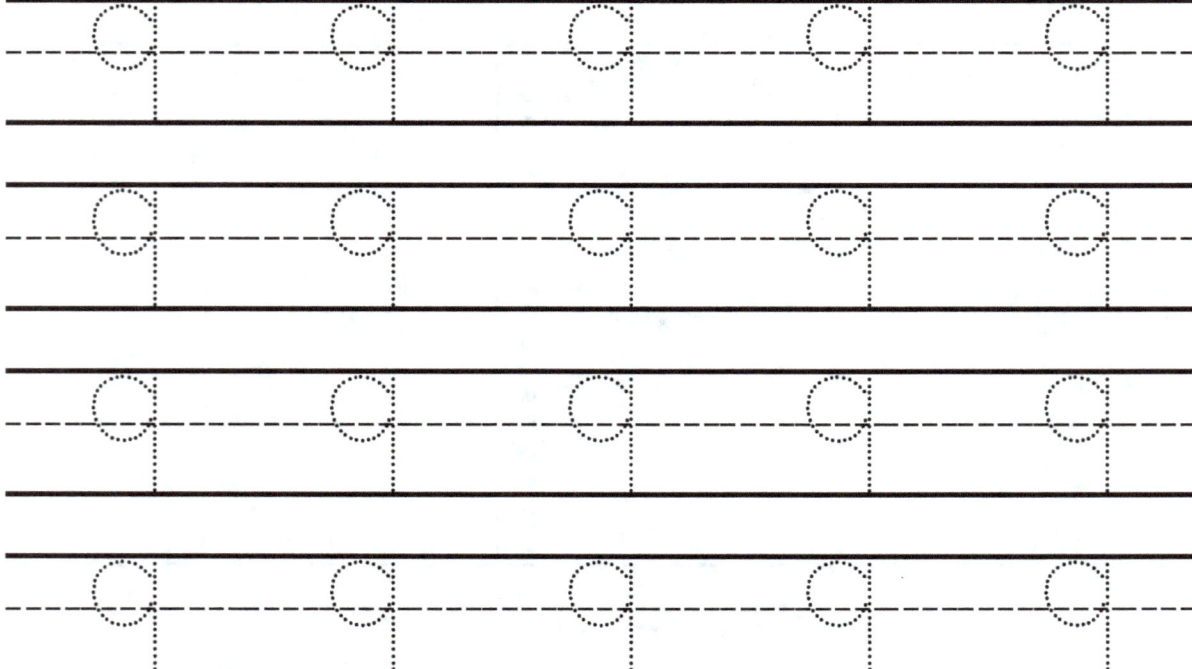

Understanding the number 10

In Words	Symbol	Count
Ten	10	

Not it's *Your* Turn!

In Words	Symbol	Count
Ten	10	1 2 3 4 5 6 7 8 9 10
Ten	10	

TRACING NUMBERS

Count the items and trace the number word.

Trace the number.

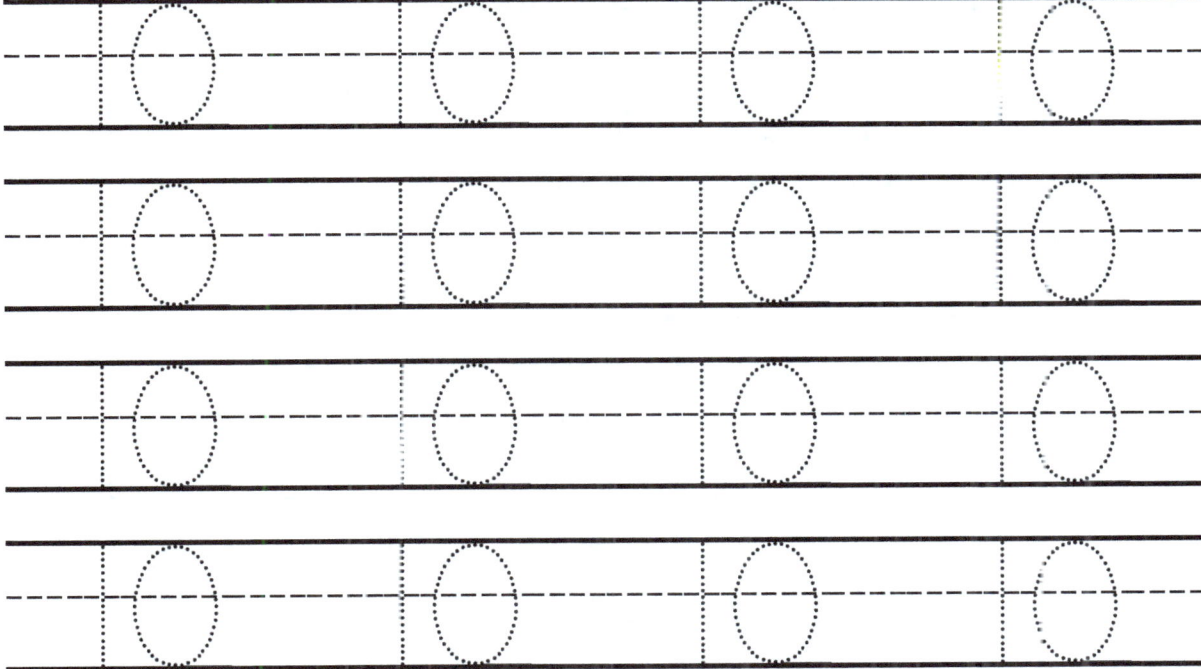

TRACING NUMBERS

Count the items and trace the number word.

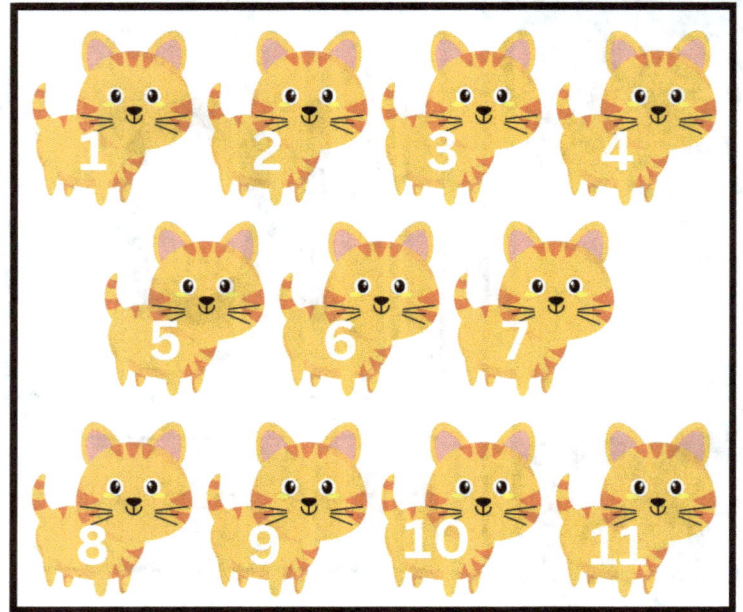
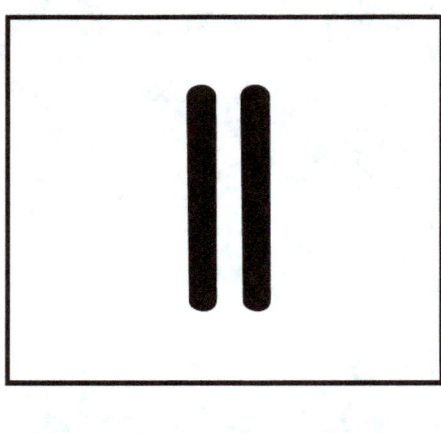

Trace the number.

TRACING NUMBERS

Count the items and trace the number word.

twelve

Trace the number.

12 12 12 12
12 12 12 12
12 12 12 12
12 12 12 12

TRACING NUMBERS

Count the items and trace the number word.

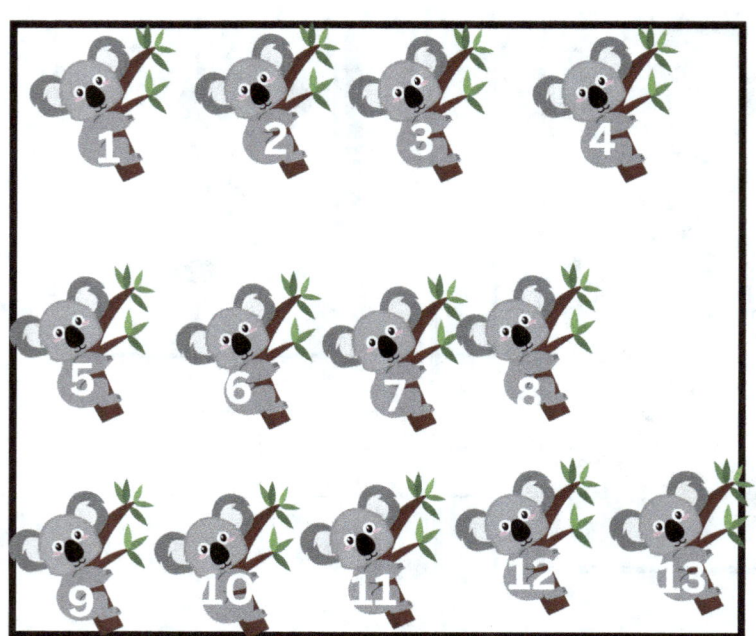

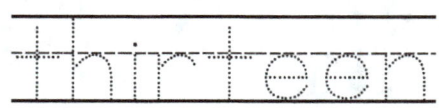

Trace the number.

TRACING NUMBERS

Count the items and trace the number word.

Trace the number.

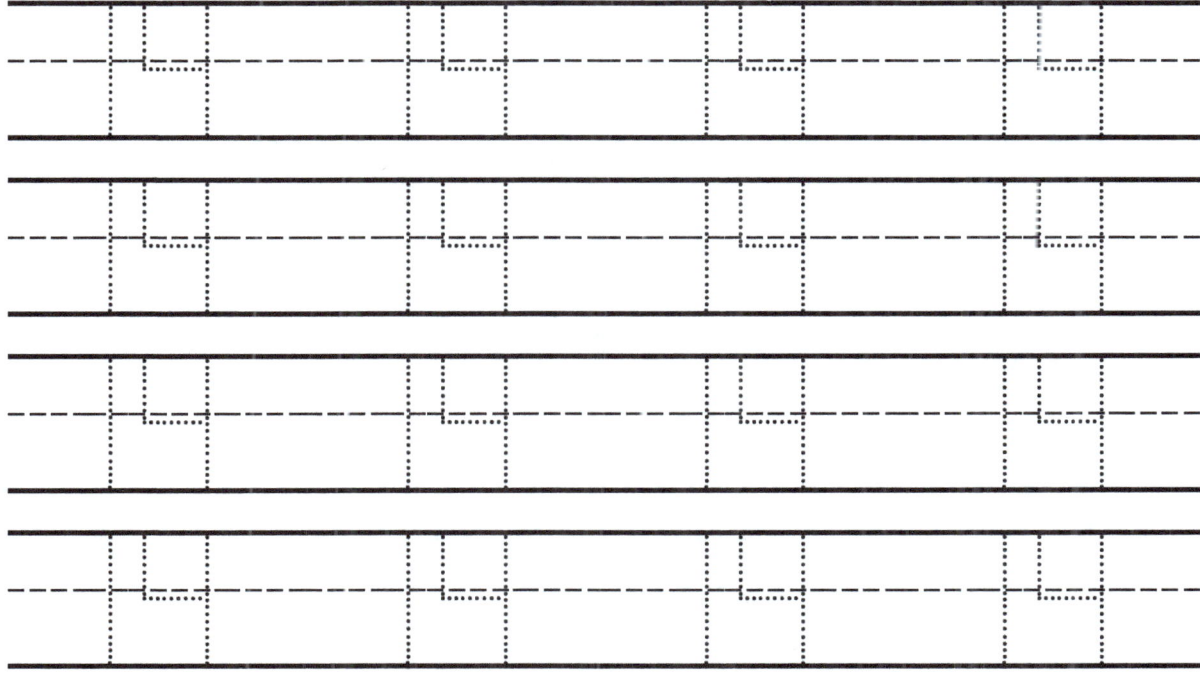

TRACING NUMBERS

Count the items and trace the number word.

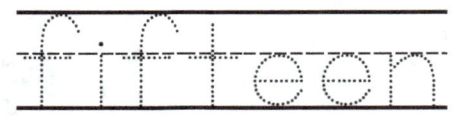

Trace the number.

15 15 15 15

15 15 15 15

15 15 15 15

15 15 15 15

TRACING NUMBERS

Count the items and trace the number word.

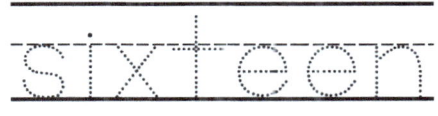

Trace the number.

TRACING NUMBERS

Count the items and trace the number word.

Trace the number.

TRACING NUMBERS

Count the items and trace the number word.

Trace the number.

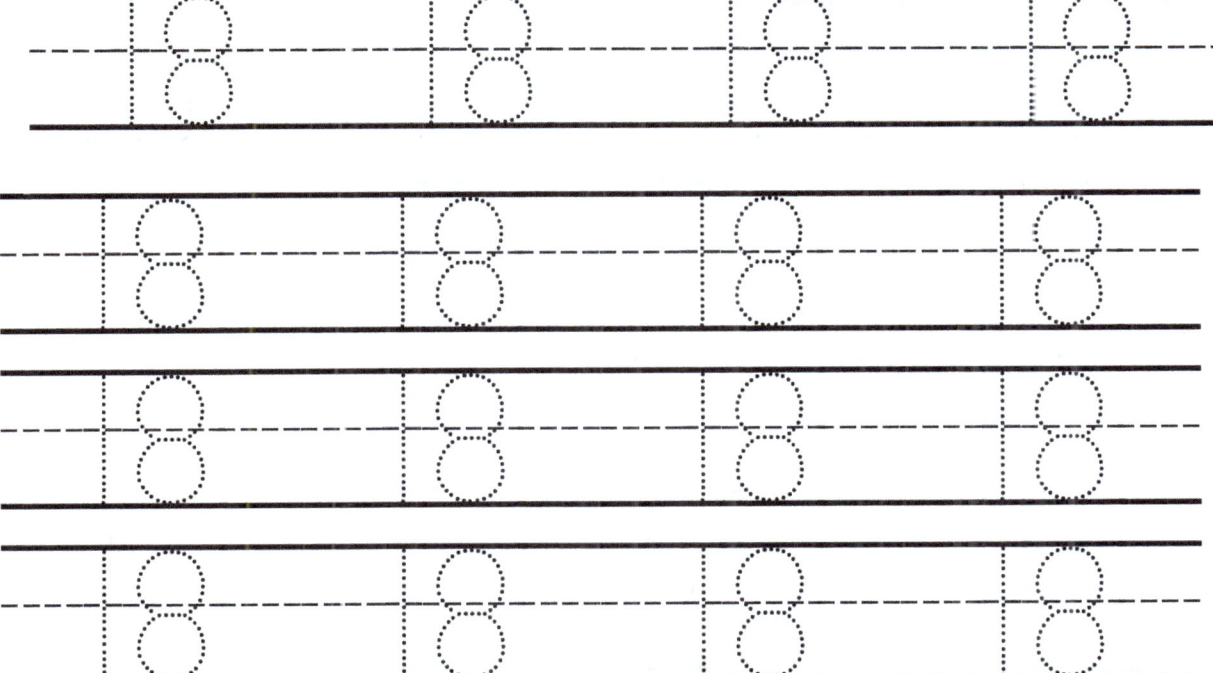

TRACING NUMBERS

Count the items and trace the number word.

nineteen

Trace the number.

19 19 19 19

19 19 19 19

19 19 19 19

19 19 19 19

TRACING NUMBERS

Count the items and trace the number word.

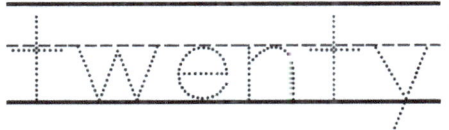

Trace the number.

Counting from 1-10

10										
9	**9**									
8	8	**8**								
7	7	7	**7**							
6	6	6	6	**6**						
5	5	5	5	5	**5**					
4	4	4	4	4	4	**4**				
3	3	3	3	3	3	3	**3**			
2	2	2	2	2	2	2	2	**2**		
1	1	1	1	1	1	1	1	1	**1**	

COUNTING FROM 1-100

Word	Number
One	01
Two	02
Three	03
Four	04
Five	05
Six	06
Seven	07
Eight	08
Nine	09

One	
Two	
Three	
Four	
Five	
Six	
Seven	
Eight	
Nine	

Word	Number
Ten	10
Eleven	11
Twelve	12
Thirteen	13
Fourteen	14
Fifteen	15
Sixteen	16
Seventeen	17
Eighteen	18
Nineteen	19

Ten	
Eleven	
Twelve	
Thirteen	
Fourteen	
Fifteen	
Sixteen	
Seventeen	
Eighteen	
Nineteen	

10	
11	
12	
13	
14	
15	
16	
17	
18	
19	

Word	Number
Twenty	20
Twenty-one	21
Twenty-two	22
Twenty-three	23
Twenty-four	24
Twenty-five	25
Twenty-six	26
Twenty-seven	27
Twenty-eight	28
Twenty-nine	29

Twenty	
Twenty-one	
Twenty-two	
Twenty - three	
Twenty-four	
Twenty-five	
Twenty-six	
Twenty-seven	
Twenty-eight	
Twenty-nine	

20	
21	
22	
23	
24	
25	
26	
27	
28	
29	

Word	Number
Thirty	30
Thirty-one	31
Thirty-two	32
Thirty-three	33
Thirty-four	34
Thirty-five	35
Thirty-six	36
Thirty-seven	37
Thirty-eight	38
Thirty-nine	39

Thirty	
Thirty-one	
Thirty-two	
Thirty - three	
Thirty-four	
Thirty-five	
Thirty-six	
Thirty-seven	
Thirty-eight	
Thirty-nine	

30	
31	
32	
33	
34	
35	
36	
37	
38	
39	

Word	Number
Forty	40
Forty-one	41
Forty-two	42
Forty-three	43
Forty-four	44
Forty-five	45
Forty-six	46
Forty-seven	47
Forty-eight	48
Forty-nine	49

Forty	
Forty-one	
Forty-two	
Forty - three	
Forty-four	
Forty-five	
Forty-six	
Forty-seven	
Forty-eight	
Forty-nine	

40

41

42

43

44

45

46

47

48

49

Word	Number
Fifty	50
Fifty-one	51
Fifty-two	52
Fifty-three	53
Fifty-four	54
Fifty-five	55
Fifty-six	56
Fifty-seven	57
Fifty-eight	58
Fifty-nine	59

Fifty	
Fifty-one	
Fifty-two	
Fifty - three	
Fifty-four	
Fifty-five	
Fifty-six	
Fifty-seven	
Fifty-eight	
Fifty-nine	

50

51

52

53

54

55

56

57

58

59

Word	Number
Sixty	60
Sixty-one	61
Sixty-two	62
Sixty-three	63
Sixty-four	64
Sixty-five	65
Sixty-six	66
Sixty-seven	67
Sixty-eight	68
Sixty-nine	69

Sixty	
Sixty-one	
Sixty-two	
Sixty - three	
Sixty-four	
Sixty-five	
Sixty-six	
Sixty-seven	
Sixty-eight	
Sixty-nine	

60	
61	
62	
63	
64	
65	
66	
67	
68	
69	

Word	Number
Seventy	70
Seventy-one	71
Seventy- two	72
Seventy-three	73
Seventy-four	74
Seventy- five	75
Seventy-six	76
Seventy-seven	77
Seventy-eight	78
Seventy-nine	79

Seventy	
Seventy-one	
Seventy-two	
Seventy - three	
Seventy-four	
Seventy-five	
Seventy-six	
Seventy-seven	
Seventy-eight	
Seventy-nine	

70	
71	
72	
73	
74	
75	
76	
77	
78	
79	

Word	Number
Eighty	**80**
Eighty-one	81
Eighty- two	82
Eighty-three	83
Eighty-four	84
Eighty- five	85
Eighty-six	86
Eighty-seven	87
Eighty-eight	88
Eighty-nine	89

Eighty	
Eighty-one	
Eighty-two	
Eighty- three	
Eighty-four	
Eighty-five	
Eighty-six	
Eighty-seven	
Eighty-eight	
Eighty-nine	

80

81

82

83

84

85

86

87

88

89

Word	Number
Ninety	90
Ninety-one	91
Ninety-two	92
Ninety-three	93
Ninety-four	94
Ninety- five	95
Ninety-six	96
Ninety-seven	97
Ninety-eight	98
Ninety-nine	99

Ninety	
Ninety-one	
Ninety-two	
Ninety - three	
Ninety-four	
Ninety-five	
NInety-six	
Ninety-seven	
Ninety-eight	
Ninety-nine	

90

91

92

93

94

95

96

97

98

99

One Hundred

100

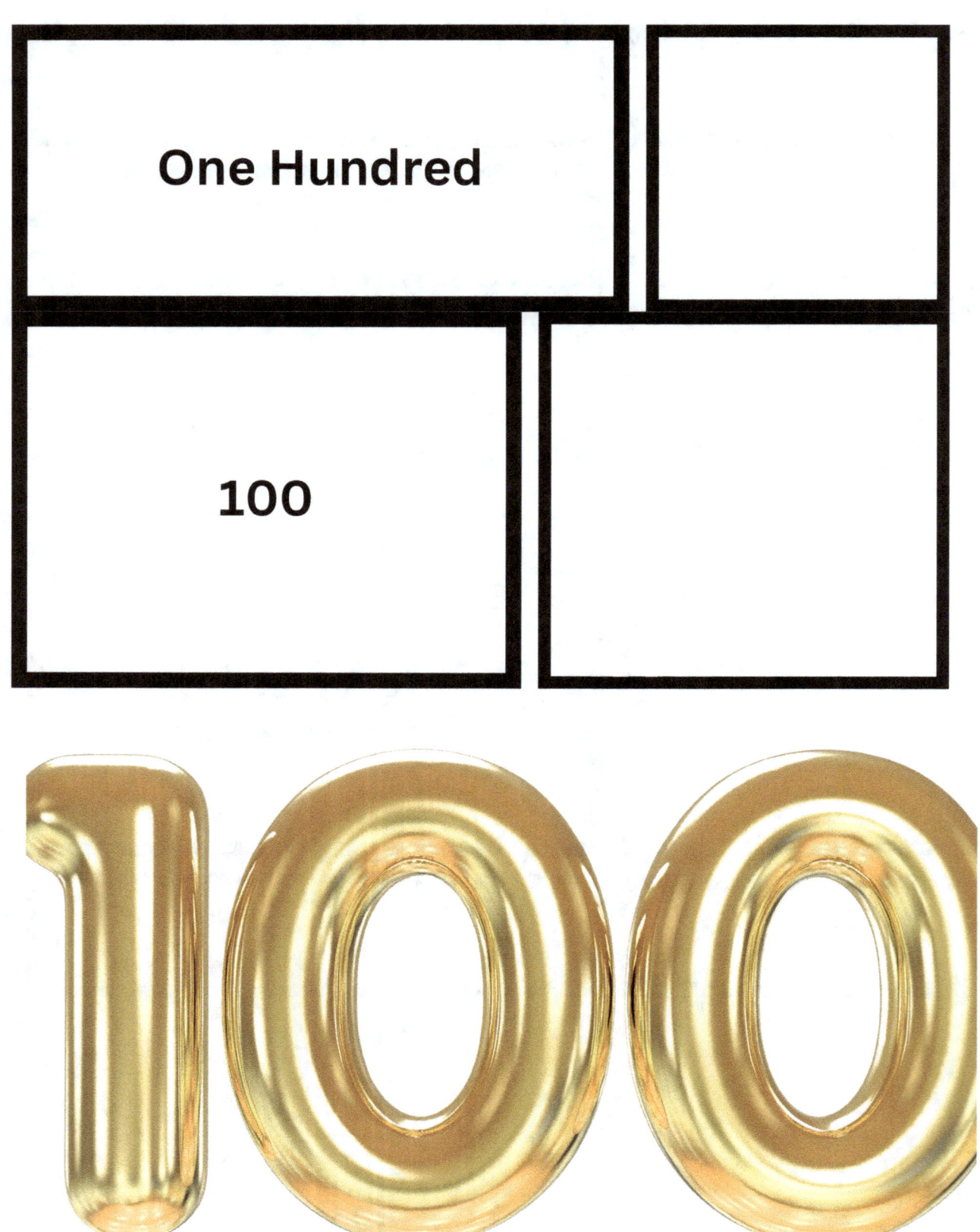

Counting 1 to 20

READ IT

1	2	3	4	5	6	7	8	9	10
11	12	13	14	15	16	17	18	19	20

TRACE IT

1	2	3	4	5	6	7	8	9	10
11	12	13	14	15	16	17	18	19	20

WRITE IT

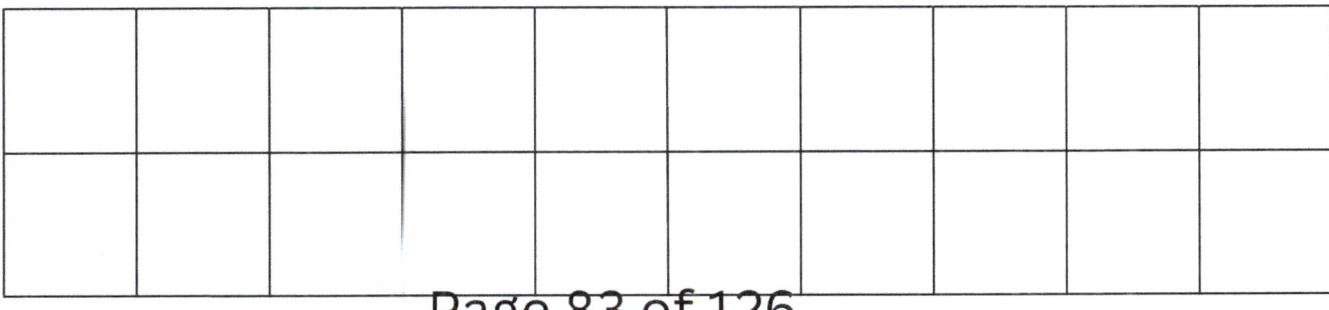

| 1 | 2 | 3 | 4 | 5 | 6 | 7 | 8 | 9 |

Count and circle the number of fruits in each box.

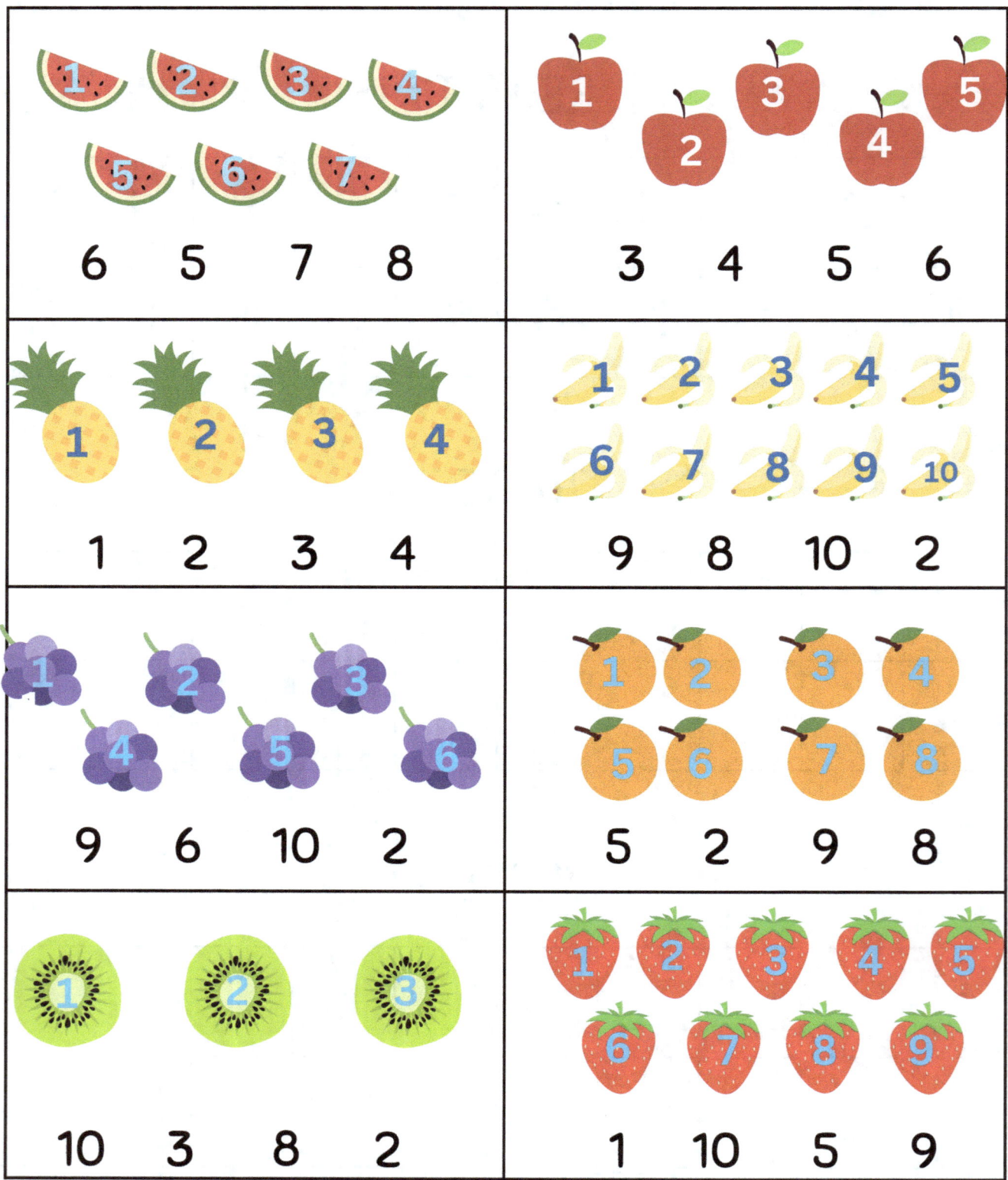

LET'S COUNT ANIMALS! HAVE AN ADULT TO CUT OUT EACH ANIMAL AND THEN COUNT

Count and write your answers in the chart below

ADDITION PROBLEMS

How many?

Count the bugs and write the numbers (1-10) in the boxes.

Use Your Math Manipulatives and your Dry Erase Board

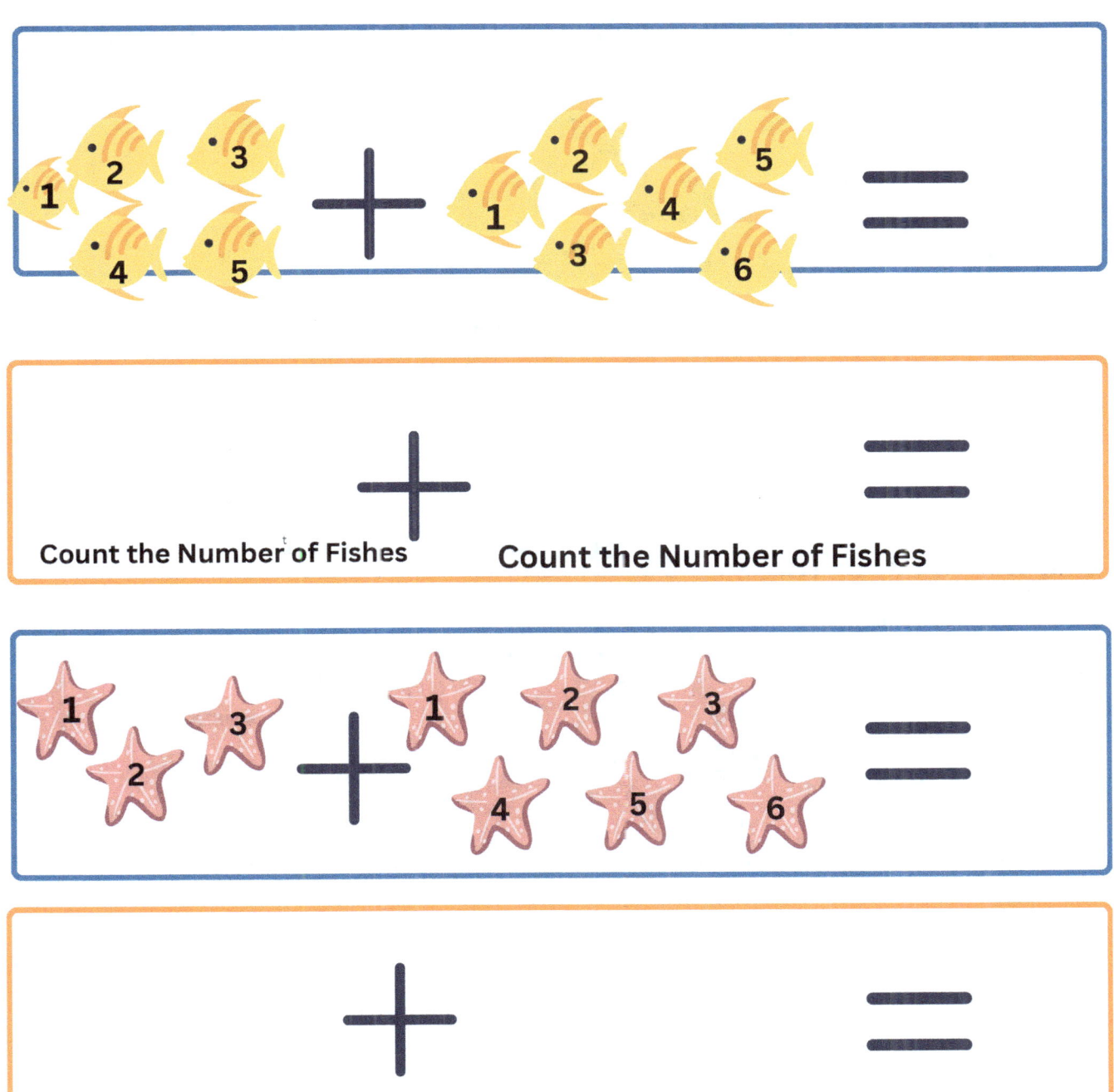

Count the Number of Fishes   Count the Number of Fishes

Count the Number of Stars   Count the Number of Stars

Use Your Math Manipulatives and your Dry Erase Sheets

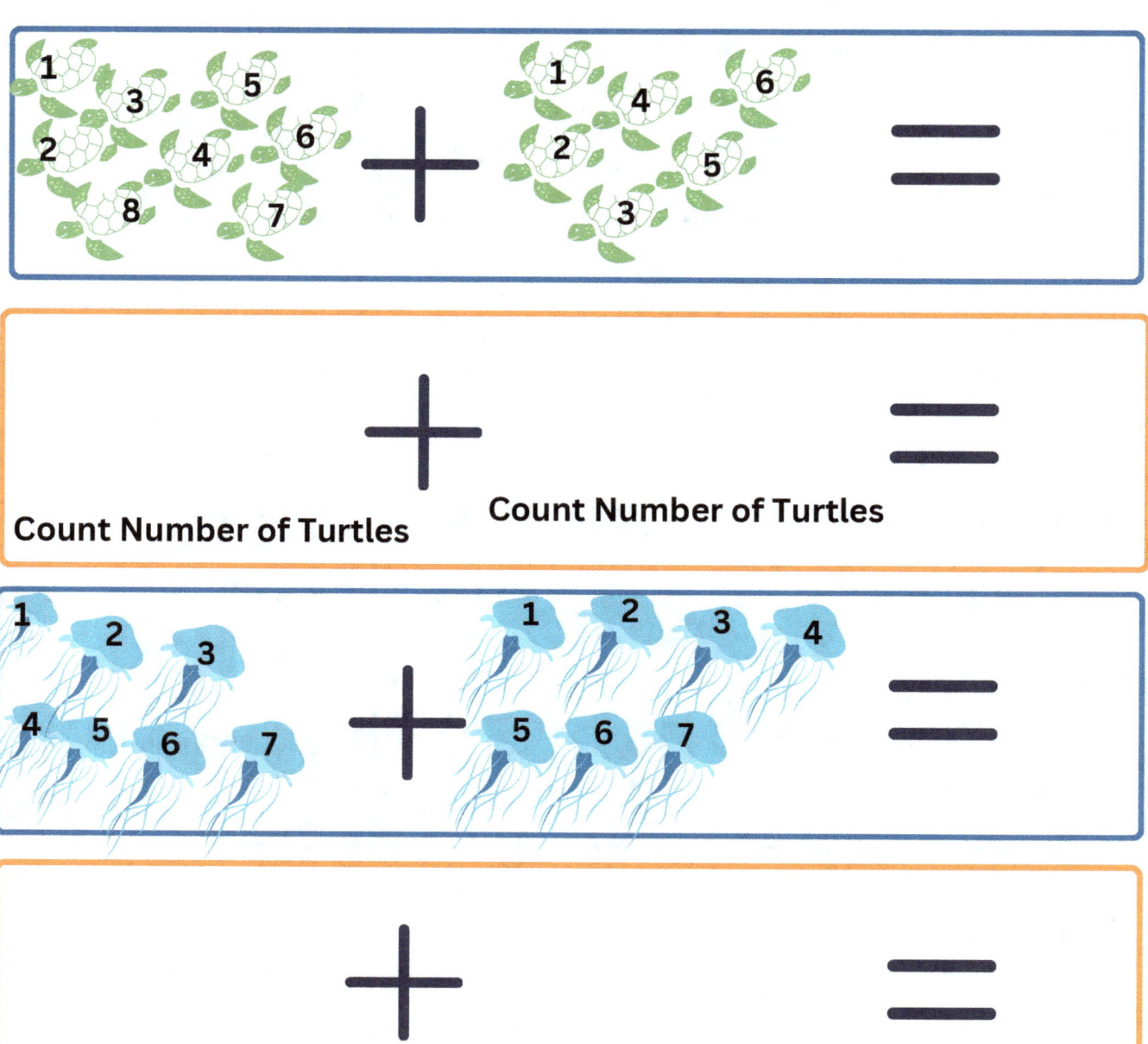

Count Number of Turtles   Count Number of Turtles

Count Number of Jelly Fishes   Count Number of Jelly Fishes

FIND THE MISSING NUMBER

USE YOUR MATH MANIPULATIVES AND YOUR DRY ERASE BOARD

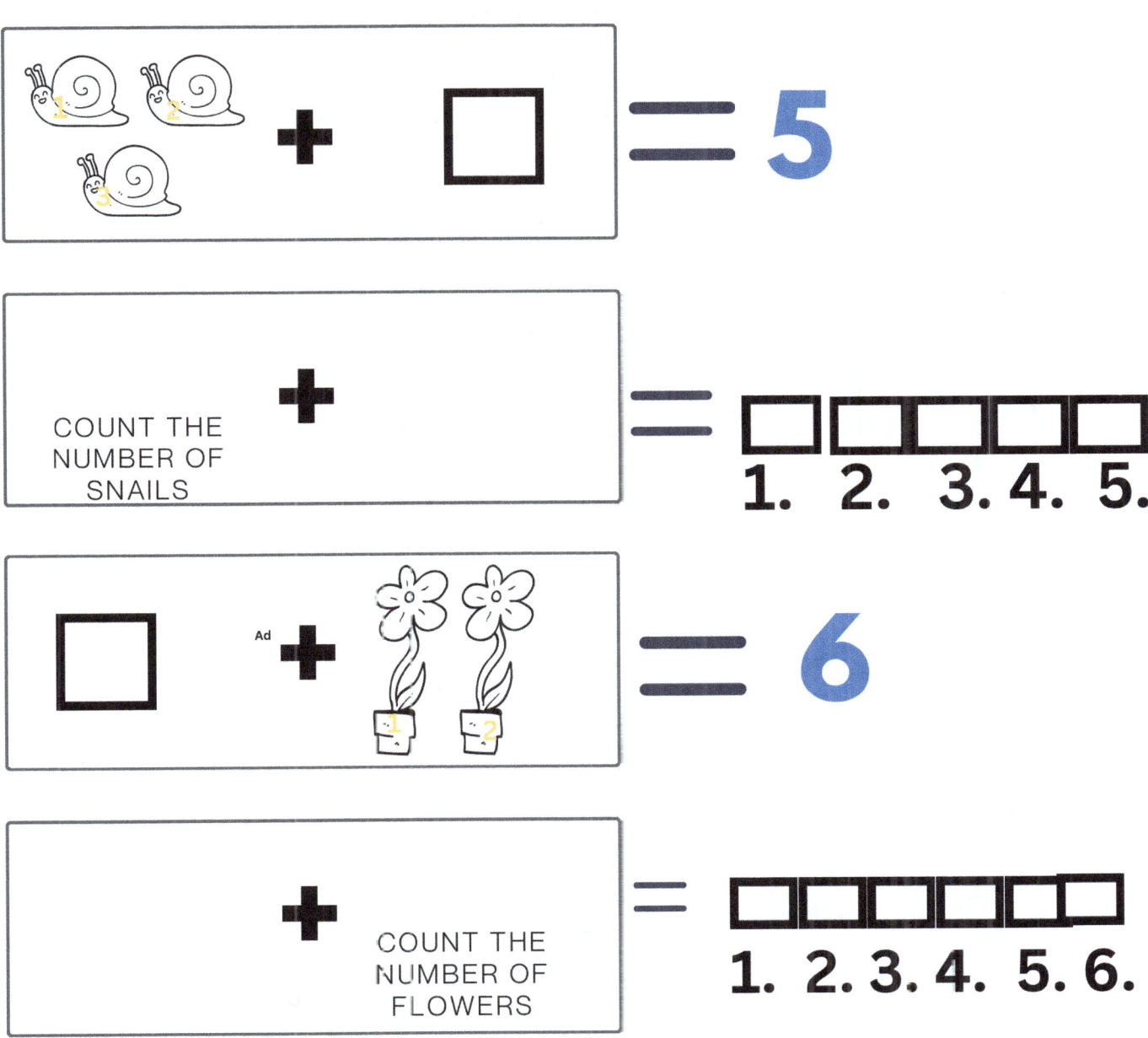

| 1 | 2 | 3 | 4 | 5 | 6 | 7 | 8 | 9 | 10 | 11 | 12 | 13 | 14 |

USE YOUR MATH MANIPULATIVES AND YOUR DRY ERASE BOARD
Fill in the missing number to find the answer!

3 + ☐ = 10

7 − ☐ = 5

11 − ☐ = 6

4 + ☐ = 10

4 + ☐ = 8

3 + ☐ = 7

12 − ☐ = 6

2 + ☐ = 9

MILLIONS ~ N TIMES 1,000,000	HUNDRED THOUSANDS ~ N TIMES 100,000	TEN THOUSANDS ~ N TIMES 10,000	THOUSANDS ~ N TIMES 1,000	HUNDREDS ~ N TIMES 100	TENS ~ N TIMES 10	ONES ~ N TIMES 1

PLACE VALUE TOOL

Million	Hunded-thousands	Ten-thousands	Thousands	Hundreds	Tens	Ones
Total						

+

PLACE VALUE
For Addition

22
+6

Hundreds	Tens	Ones
	2	2
Plus	0	6
Answer	2	8

New Problem- 2 Digit

	Tens	Ones
Plus	4	6
	5	2
Answer	9	8

PLACE VALUE

For 3-Digit Addition Tool

Hundreds	Tens	Ones
Answer		

New 2 Digit Addition Tool

	Tens	Ones
Plus		
Answer		

Million	Hunded-thousands	Ten-thousands	Thousands	Hundreds	Tens	Ones
				6	4	3
				1	3	2
Total						
+				7	7	5

or
700+ 70+5 = 775
Or 7 hundreds
7 tens

3 Digit Addition- Use You Place Value Chart

Solve each addition equation.

643 +132 ─── 775	223 +147 ───	361 +749 ───	146 +621 ───
356 +951 ───	241 +614 ───	238 +132 ───	643 +132 ───
219 +852 ───	358 +232 ───	178 +335 ───	727 +152 ───
382 +266 ───	711 +244 ───	455 +782 ───	468 +481 ───

PLACE VALUE
Addition Tool- with regrouping

$$\begin{array}{r}314\\5\\+\ 23\end{array}$$

	Hundreds	Tens	Ones
	3	1	4
		0	5
Plus		2	3
Subtotal	3	3	***12***
	3	3(adding)	1 ten 2 ones Need to regroup
		1	2
Final Answer	**3**	**4**	**2**

Addition with Regrouping
Use You Place Value Chart

314 5 + 23	200 40 + 22	282 12 + 14	88 7 + 165
414 13 + 412	57 365 + 22	63 215 + 19	174 22 + 45
112 63 + 320	22 95 + 122	36 152 + 91	471 122 + 54

PLACE VALUE

For 3-Digit Addition Tool

$$276$$
$$+\ 81$$

	Hundreds	Tens	Ones
	2	7	6
Plus		8	1
		15 times 10= 150 **1 hundred 5 tens**	7
	2	0	7
Subtotal	1	5	0
Answer	3	5	7

```
 276      297       81      199      104
+ 81     + 34     + 55     +177     + 217
----     ----     ----     ----     ----


                                      87
 275      57       66      371
+294     +42     + 37     +278      + 56
----     ---     ----     ----      ----


 345      18      493
+ 155    + 18    +398
-----    ----    ----
```

SUBTRACTION PROBLEMS

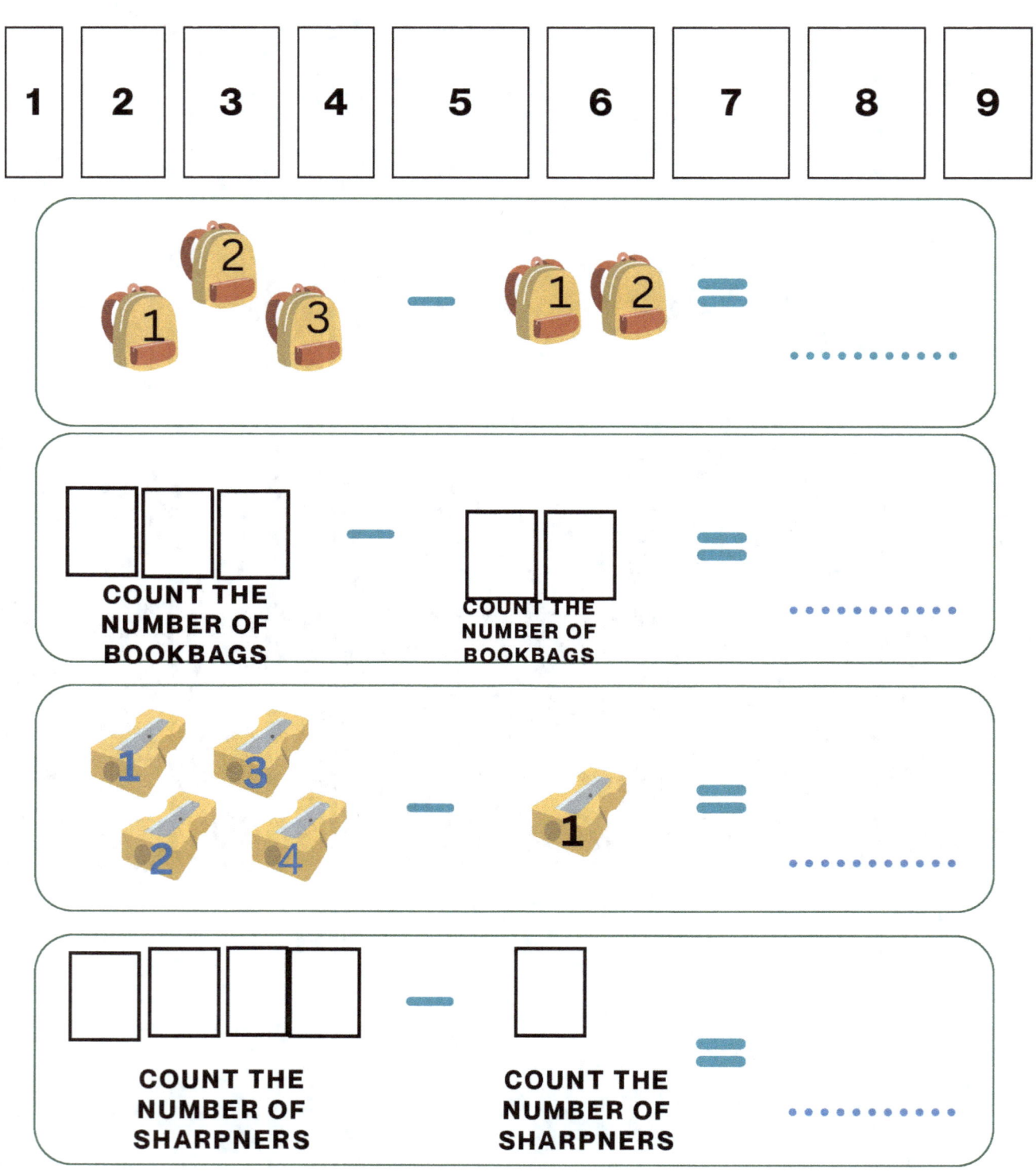

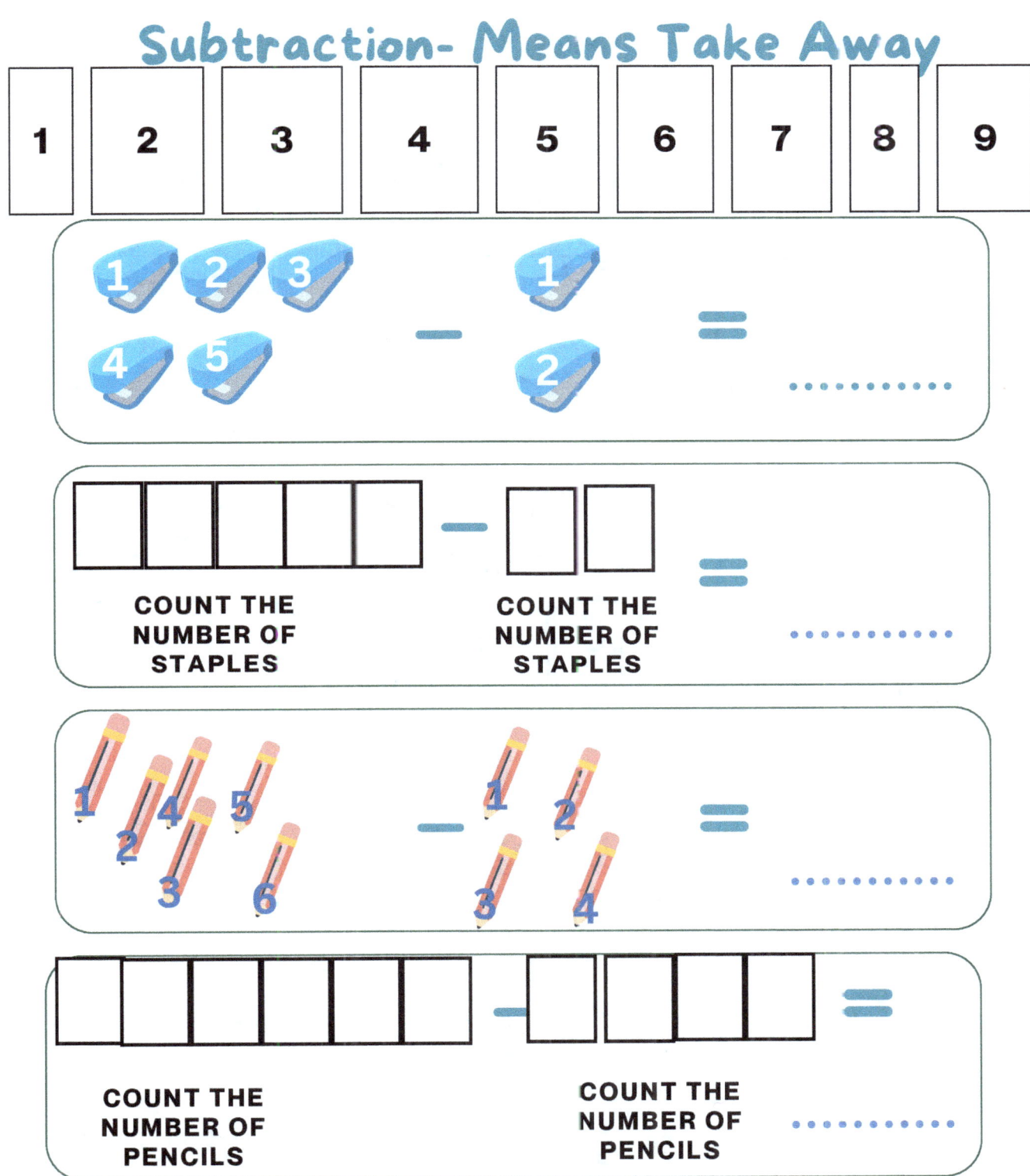

Subtraction- Take Away

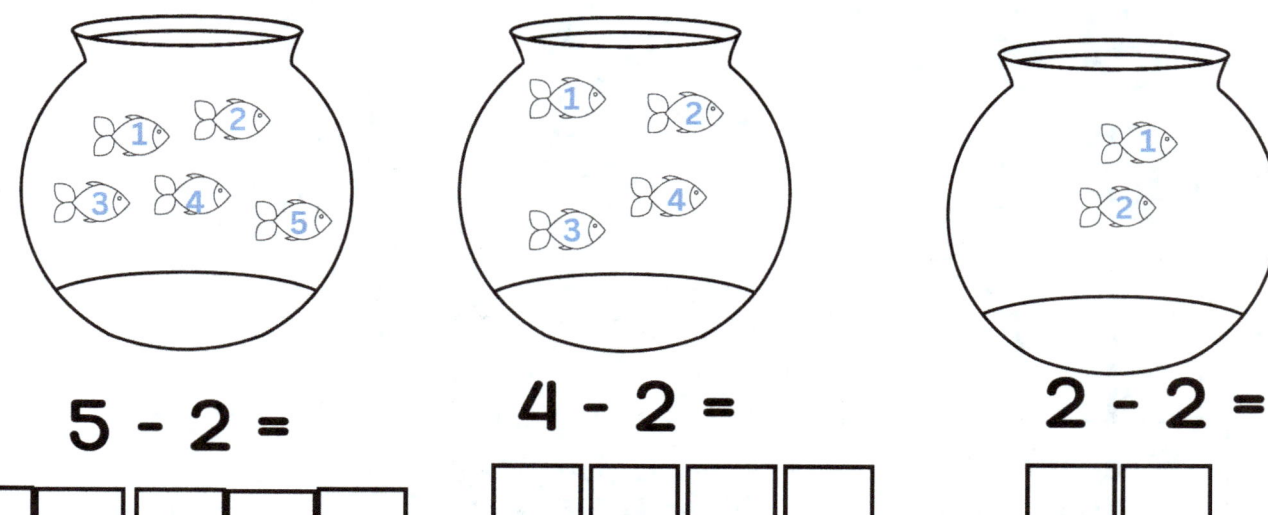

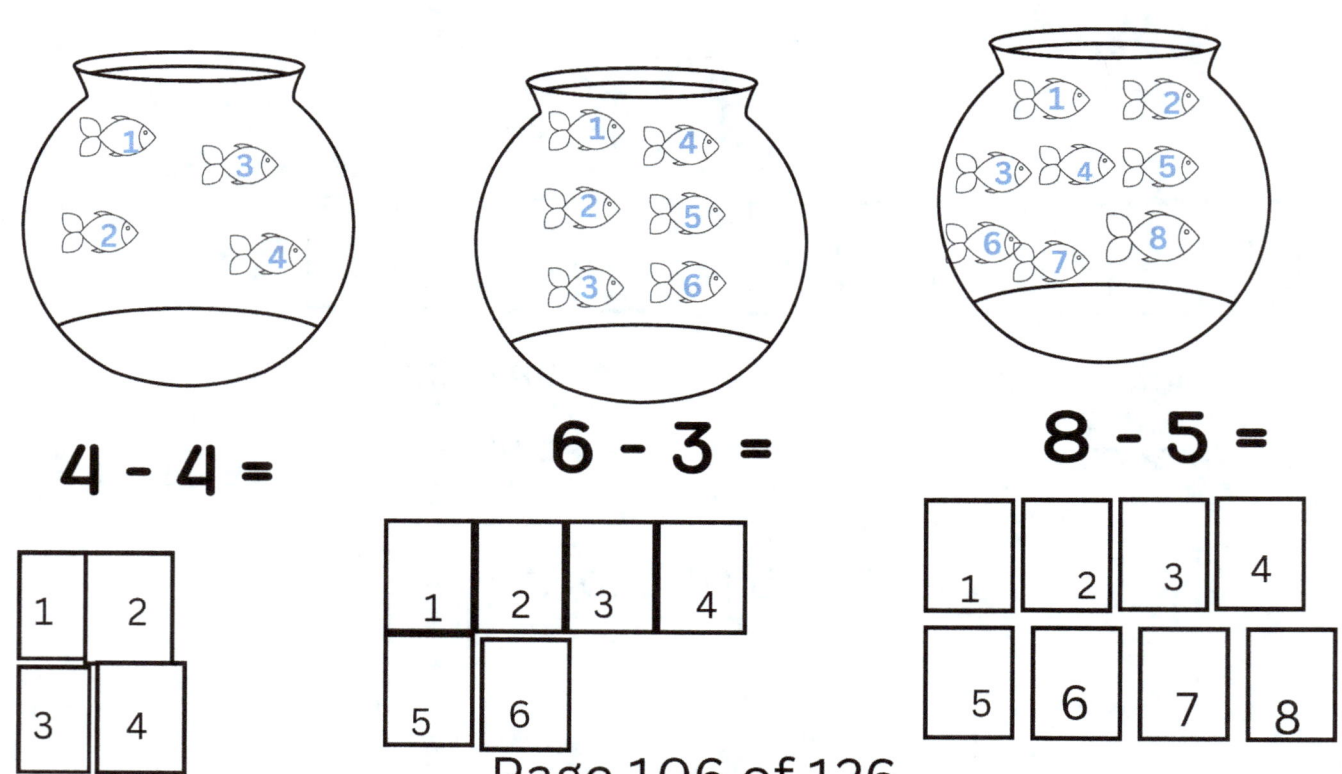

Million	Hunded-thousands	Ten-thousands	Thousands	Hundreds	Tens	Ones
Total						

+

Million	Hunded-thousands	Ten-thousands	Thousands	Hundreds	Tens	Ones
+						
Total						

PLACE VALUE

For 2-Digit Subtraction Tool - Without Borrowing

87
- 36

Hundreds	Tens	Ones
	8	7
Minus	3	6
Answer	5	1

PLACE VALUE

For 3-Digit Addition Tool

6 8 5

- <u>2 7 2</u>

	Hundreds	Tens	Ones
	6	8	5
Minus	2	7	2
Answer	4	1	3

PLACE VALUE

For 3-Digit Addition Tool

$$571$$
$$\underline{-163}$$

	Hundreds	Tens	Ones
	5	7̷ 6	1 (need to borrow 10) 11
Minus	1	6	3
Answer	4	0	8

PLACE VALUE

For 3-Digit Addition Tool

600

-263

	Hundreds	₇Tens	Ones
	~~6~~ 5	~~0~~ 9 (Need to borrow 10	0 (need to borrow 10) 10
Minus	2	6	3
Answer	3	3	7

Use Your Math Tools

80	51	75	48
− 36	− 34	− 19	− 19

45	65	70	41
− 19	− 27	− 38	− 19

50	89	74	75
− 26	− 38	− 26	− 16

```
  326      87      73     978     503
-  85    - 59    - 25    -229    -124
_____    ____    ____    ____    ____


  265     97      81     300     433
- 194   - 23    - 56    - 92    -182
_____   ____    ____    ____    ____


  846    524     600     741     800
- 508   -439    -143    -326    -104
_____   ____    ____    ____    ____
```

	Hundreds	Tens	Ones
	9	7 6	8 (need to borrow 10) 18
Minus	2	2	9
Answer	7	4	9

PLACE VALUE PROBLEMS

MILLIONS ~ N TIMES 1,000,000	HUNDRED THOUSANDS ~ N TIMES 100,000	TEN THOUSANDS ~ N TIMES 10,000	THOUSANDS ~ N TIMES 1,000	HUNDREDS ~ N TIMES 100	TENS ~ N TIMES 10	ONES ~ N TIMES 1

PLACE VALUE TOOL

Million	Hunded-thousands	Ten-thousands	Thousands	Hundreds	Tens	Ones
+						
			Total			

PLACE VALUE

Number	Hundreds N x 100	Tens N X 10	Ones N x 1
362 Standard Form	3	6	2
Expanded Notation	300	60	2
Word Form: Three Hundred Sixty-Two			
Number	Hundreds N x 100	Tens N X 10	Ones N x 1
109 Standard Form	1	0	9
Expanded Notation	100	0	9
Word Form : One hundred Nine			
Number	Hundreds N x 100	Tens N X 10	Ones N x 1
38 Standard Form		3	8
Expanded Notation		30	8

EXPANDED NOTATION

Expanded notation is a way of representing a number by breaking it down into its individual place values.

ten thousands	thousands	hundreds	tens	ones
3	2	9	1	6

3 · 10,000
2 · 1,000
9 · 100
1 · 10
6 · 1

30,000 + 2,000 + 900 + 10 + 6 = 32,916

Standard Notation = 32,916

50,000+ 5,000+ 300+ 40+ 1

	Ten-thousands		Thousands	Hundreds	Tens	Ones
		5	0	0	0	0
			5	0	0	0
				3	0	0
+					4	0
						1
		Total				
		5	5	3	4	1

PLACE VALUE

EXPANDED FORM STANDARD FORM

Convert each number from expanded form to standard from.

40,000+ 5,000+ 300+ 40+ 1	
9,000,000 + 900,000 + 70,000 + 8,000 + 400 + 10+ 4	
30,000+ 6,000+ 800+ 20+ 2	
6,000,000 + 300,000 + 80,000 + 2,000 + 600 + 80+ 9	
500,000 + 50,000 + 9,000 + 700 + 20 + 7	
600,000 + 90,000 + 3,000 + 800 + 20 + 9	
8,000,000 + 300,000 + 60,000 + 2,000 + 700 + 70+ 1	
800,000 + 30,000 + 9,000 + 300 + 60 + 6	

MILLIONS ~ N TIMES 1,000,000	HUNDRED THOUSANDS ~ N TIMES 100,000	TEN THOUSANDS ~ N TIMES 10,000	THOUSANDS ~ N TIMES 1,000	HUNDREDS ~ N TIMES 100	TENS ~ N TIMES 10	ONES ~ N TIMES 1

PLACE VALUE TOOL

PLACE VALUE

STANDARD FORM ➡ EXPANDED FORM

Convert each number from standard from to expanded form.

7,218,181	
6,857,921	
5,973,963	
3,242,142	
344,332	
432,677	
2,346,722	
23,552,211	
78,322	
2,355,696	

Use PLACE VALUE AND VALUE Tool

Write the place value and value of the underlined digit.

	Place Value	Value
48<u>6</u>	tens	80
92<u> </u>		
<u>5</u>71		
81<u>2</u>		
1 6<u>9</u>3		
4 <u>2</u>57		
2 915<u> </u>		
5 9<u>3</u>4		
3 99<u>1</u>		
<u>7</u> 168		

Keep Dreaming for a better tomorrow. Our students are depending on adults to work together on their behalf.

ABOUT

DR. JANNELL PEARSON-CAMPBELL, ED.D

Dr. Jannell Pearson-Campbell is a 22-year educational leader and practitioner with a career dedicated to serving diverse learners. Dr. Pearson-Campbell has been a teacher and administrative leader in urban and suburban districts, where she has developed leaders in diversity, equity, and inclusion in academic settings.

Drawing from her experience in teaching math, science, English language instruction, and Special Education in Boston and Framingham Public Schools, Dr. Pearson-Campbell has successfully used data, protocols, and policies to drive instruction for all students.

Dr. Pearson-Campbell is currently creating math resource guides and creative writing journals for teachers, students and community members to use in order to close the achievement gap in math.

Dr. Jannell Pearson-Campbell can be reached at Dr.JPCunlimited@gmail.com. and drjpcdownloads.etsy.com

www.ingramcontent.com/pod-product-compliance
Lightning Source LLC
Chambersburg PA
CBHW062108220526
45471CB00010B/3649